Central Rockies MAMMALS

Central Rockies MAMMALS

LUMINOUS
COMPOSITIONS

Published by
Luminous Compositions
P.O. Box 2112, Banff, Alberta, Canada T0L 0C0
e-mail: luminous@agt.net

Canadian Cataloguing in Publication Data

Marriott, John, 1969-
Central Rockies Mammals

(A 'Pack-it' Pocket Guide)
ISBN 0-9694438-7-0

1. Mammals—Rocky Mountains, Canadian (B.C. and Alta.).
I. Title. II. Series.

QL221.R6M37 1997 599'.09711 C97-900385-7

Printed on recycled paper and bound by
Quality Color Press, Edmonton, Alberta, Canada

Front cover photograph: Black bear eating dandelions. (John Marriott)
Back cover photograph: Coyote in winter. (John Marriott)

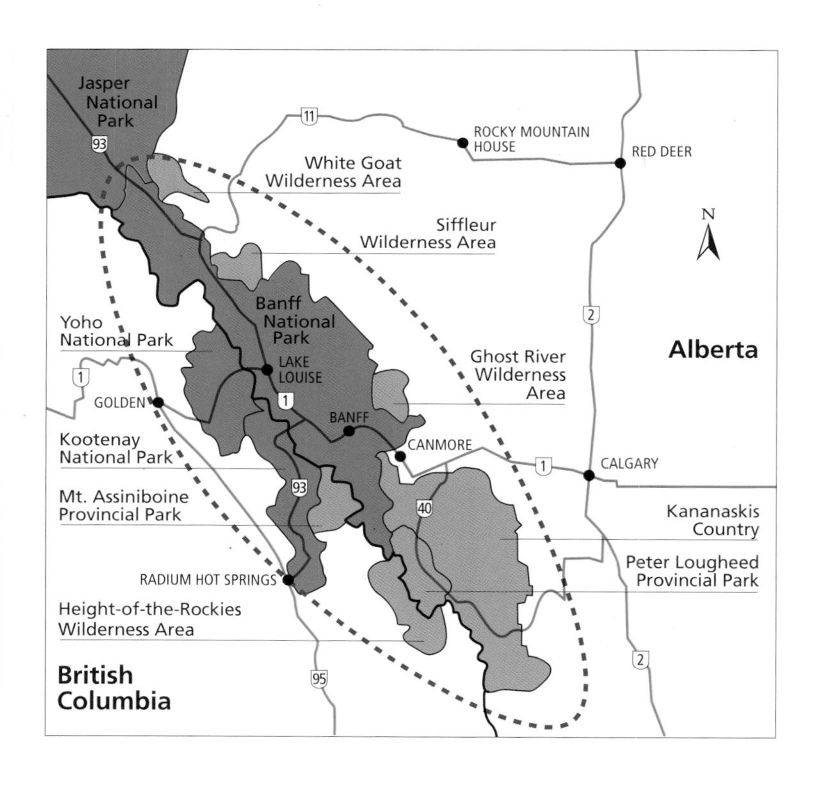

*The **Central Rockies Ecosystem** is the area inside the blue dashed line.*

The first time I came face-to-face with a wolf was in June, 1993, near a peaceful stretch of the Bow River beneath Castle Mountain in Banff National Park. Driving beside the river on a trip from Banff to Lake Louise, I spotted a large wolf--grey in colour--walking along the far shore. I pulled off the highway into a small parking lot and stood beside my car as the wolf slowly approached until it was directly opposite me, fifty metres away. To my astonishment, the wolf looked right at me, then sat down on its haunches and gazed into the forest behind it.

After a moment, the wolf disappeared into the forest, then re-emerged fifteen seconds later further down the river, still peering intently into the trees. I moved out to the river's edge just in time to see a cow elk crash through the vegetation two hundred metres upstream and plunge into the icy water. Following close on her heels were two different smaller wolves, one dark grey, the other tawny brown. The big wolf I had been watching joined in the chase, but the elk soon powered away from the wolves, leaving them to return to shore where they quietly blended back into the forest.

The wolves I observed that afternoon were members of the Bow Valley pack, which frequents the Bow Valley in the heart of the Central Rockies Ecosystem (see map opposite and glossary, p. 93). The wolves share the valley with a number of large mammals, including grizzly bears, black bears, cougars, elk, bighorn sheep and deer, and with a host of smaller mammals, such as beavers, flying squirrels and deer mice. The Bow Valley, like many of the valleys in the Central Rockies, is also shared with humans, resulting in a curious mix of dens, burrows and human developments. Wildlife corridors often traverse highways and railways, while backcountry trails have humans on them one minute, grizzly bears the next.

John Marriott

This delicate balance provides human visitors and residents alike with a unique opportunity to view many of North America's most fascinating animals in these mountains. It also requires that we take responsibility to ensure that the Central Rockies remain a wild environment for the diversity of creatures found here.

The balance can shift quickly in the Central Rockies, making this area a hotbed for wildlife research – in particular, for the study of human-animal interaction. Current studies include research on grizzly bears, black bears, wolves, cougars, lynx, wolverines, moose, elk and mountain goats. There are positive stories, such as the apparently successful reintroduction of mountain goats into parts of Kananaskis Country and the natural return of wolves to Banff National Park in the early 1980s.

Unfortunately, there are also discouraging findings – black bears are a threatened species in Banff National Park, while today more wolves in the Central Rockies are killed by hunters, trains and vehicles than by all other causes of death combined.

There is still a very good chance of spotting wildlife when hiking or driving in the Central Rockies, particularly in the spring and fall when there are fewer visitors in the area, and at dawn and dusk when

John Marriott

the animals are most active. A good rule of thumb to abide by when viewing wildlife is to give the animal/s enough space to not disturb them or put yourself in danger. Stay at least 50 metres (165 ft) away and use a telephoto lens or a pair of binoculars to photograph or watch wildlife from a safe distance. For bears or other aggressive or unpredictable wildlife, including elk with calves, remain in a vehicle or vacate the area immediately if you are on a bike or on foot. Each year, people are injured by elk, deer, bears and even ground squirrels, apparently unaware that enjoying a safe wildlife viewing experience is as simple as leaving the animals room to do their thing without feeling threatened.

Over a decade ago, bears flocked to the dumps in Banff and Lake Louise. Today, thanks to proper garbage disposal, the bears have returned to the wilds, although not without incident. "Goosebear," a young black bear nicknamed for his love of gooseberries, was fed human food a number of times several summers ago, leading to officials shooting him. "Goosebear's" skin is now used at public events in Banff National Park as a vivid reminder of the dangers of feeding wildlife – it is not only illegal to feed an animal in the Central Rockies Ecosystem, it also often results in the death of the animal.

Terry Berezan

This compact book, *Central Rockies Mammals*, introduces 38 of the most common mammals found in the Central Rockies Ecosystem. Like all mammals, including humans, these species share common traits, including having highly developed brains and senses, being warm-blooded, having hair and having nipples capable of producing milk.

This book describes, and illustrates with colour photographs, the distinguishing characteristics of each species, including body size, weight, colour, and habitat preference, as well as some of the more intriguing behavioural attributes of each mammal.

The mammals are organized by taxonomic order, beginning with the carnivores, followed by ungulates, rodents, and pikas, rabbits and hares. Within each order, species are grouped according to family – for example, the carnivores are presented beginning with the mammals in the bear family, followed by the wild dogs, the cats and the weasels. Infrequent mammalian visitors to the Central Rockies, or mammals that are rarely seen, such as the northern flying squirrel, are covered in the Appendix beginning on page 90.

The aim of *Central Rockies Mammals* is to provide a keen understanding and appreciation of the wildlife that helps make this area one of the last great wilderness areas left on earth. Learning to enjoy and value wildlife is a first step to making a difference in the environment and guaranteeing that the animals will be here for future generations to watch and appreciate. I hope that you too may someday come across the tracks of a great grizzly, spend an evening camping in the company of a porcupine, or come face-to-face with a wolf.

Acknowledgements

I would like to extend my gratitude to Mike Potter for the opportunity and encouragement to produce this book. Mike also went miles providing editing, proofreading and advice.

Thanks must again go to Mike, as well as to Jeff Waugh, Al Williams and Brian Patton, for inspiring me and helping me to pursue my interests in wildlife photography.

I am grateful to Terry Berezan, Doug Leighton, Pat Morrow, Mike Potter, Paul Smith, Jeff Waugh, Al Williams and Terry Willis for their excellent photographs.

Denise Lemaster, Duane Beazer and Mike Potter did the design and pre-press. Jacob Reichbart created the original template for the "Pack-it" Pocket Guide series. Bob Doull provided essential support.

A special thank you to my parents and family for letting me pay them to watch the other side of the road for animals each time we drove through the Rockies years ago. Twenty-five cents a bear was well worth it to ensure I didn't miss anything!

And finally, I want to thank my wife Christine, who helped in the field and at the computer, giving me the support I needed to complete this book.

This book is dedicated to everyone who has felt a surge of adrenaline at the sight of a wild animal.

Grizzly Bear
Ursus arctos

John Marriott

I had my first sighting of the grizzly bear referred to as "Number Twenty-two" in 1993, while driving along Highway 40 in Kananaskis Country. At the time, she was a small young female with a distinctive beautiful blond coat and a curious regard for humans. We watched each other for over an hour as she dug on the nearby slopes in search of glacier lily bulbs.

I have been fortunate enough to watch this bear once or twice a year since that time as she has matured into a full-grown adult, and in September, 1996, I watched her and her first two young cubs (one pictured in the photo above) romp in the snow.

Grizzly bear "Twenty-two" is one of over twenty radio-collared bears being studied in the Eastern Slopes Grizzly Bear Project, a concerted effort to determine the status of grizzly bears in the Central Rockies and to ensure a long-term future for them.

Grizzly bears require huge tracts of undisturbed wilderness to survive. While a sow with two cubs like "Number Twenty-two" may have a home range of 500 square kilometres, adult males can have territories in excess of 3000 square kilometres.

Paul Smith

Adult grizzlies in the Canadian Rockies are smaller than their counterparts in Alaska, which enjoy an annual salmon feast. A full-grown boar usually weighs between 135 and 315 kg (300-700 lb) and is almost 2 m (6 ft) long. Females and subadults are considerably smaller.

Grizzly bears come in a variety of colours, including brown, black and blond. Many grizzlies have dark fur with a lighter end on each hair, making the bear's coat appear grizzled or silver-tipped.

One of the features that best distinguishes a grizzly from a black bear is the large hump of muscle visible on its back. Grizzlies acquire this mass of shoulder muscle at a young age from digging for ground squirrels, marmots, flower bulbs and plant roots. The claws of grizzlies, longer and more visible than those of black bears, assist them in their excavations.

One other feature that distinguishes a grizzly is its round, dish-shaped face. A black bear has a Roman profile, and a narrow face when viewed straight on. Body size is not a good indicator, as a big black bear can be as large as a young grizzly.

Grizzly Bear

Ursus arctos

Terry Willis

Grizzly bears are omnivores, and will eat meat or vegetation. They are not adept predators, but do occasionally bring down elk calves and other young ungulates in the spring. They will also scavenge road kills or carrion, or expend energy digging up hibernating ground squirrels or marmots. However, the bulk of a grizzly bear's diet is made up of grasses, alpine plant roots, flower bulbs, berries and other vegetarian fare.

Bear attacks on humans are not common in the Central Rockies, and grizzlies rarely attack unless surprised or threatened. Making noise while hiking and keeping a clean camp, as described in *Bear Attacks: Their Causes and Avoidance* (Herrero, 1985), are two key steps towards avoiding aggressive encounters with these magnificent animals.

Like black bears, grizzly bears do not truly hibernate, but go into a deep slumber for six months of the year from late November to early May. During this time, pregnant grizzly sows give birth to one to three cubs, and the cubs remain with their mother for two more winters before setting out to establish their own home range.

John Marriott

Past studies have indicated that grizzlies in the Central Rockies rarely die naturally; they are killed by hunters, poachers, cars, trucks, trains and wildlife managers. This high mortality, combined with a low reproductive rate and a shortage of suitable bear terrain (a large percentage of the ecosystem is ice and rock, and the best bear habitat is in the valley bottoms where our roads and towns are), has put the Central Rockies grizzly population in a precarious position.

Efforts to maintain a healthy, viable number of grizzlies, such as setting aside new protected lands and temporarily closing areas to human use (to give the bears room to move without disturbance), must be supported if we are to continue to have these symbols of the Canadian wilderness roaming free.

Black Bear

Ursus americanus

John Marriott

Black bears are common throughout the montane (low elevation) valleys of the Central Rockies. They range in colour from jet black to cinnamon brown to blond, and often have white Vs or other markings on their chests. The males, called boars, are considerably larger than the females (sows) and can be up to 1.7 m (5.5 ft) in length and 150 kg (330 lb) in weight.

The diet of most black bears includes dandelions, grasses, other plants, insects and berries. However, given the chance, a black bear will eat almost anything. I once watched a sleek brown bruin sneak up to the edge of the Trans-Canada Highway in Banff National Park and drag a deer carcass to the forest's edge. After a brief pause to catch its breath, the bear began feasting and became completely involved in its prize, seemingly forgetting that a major highway was just metres away.

Black bears will and often do eat human food, but feeding them is illegal in the Rockies because a fed bear almost always ends up being a dead bear. Inevitably, a bear that becomes used to human handouts or garbage will eventually be hit by a vehicle on the roads that it frequents, or will be killed by game officials for damaging human property.

Besides humans, the enemies of black bears include cougars, wolves and grizzly bears. Intelligent and equipped with highly developed senses, black bears are also athletic and will climb clear of threatening situations up a sturdy tree.

The same short claws that allow a black bear to climb are one of its distinguishing features when compared to a grizzly bear. Grizzlies have long claws that protrude well beyond the front of the foot and are used for digging, while a black bear's claws are more similar to the short claws on a dog. Black bears also do not have the shoulder hump of muscle that grizzlies have, and they have narrower facial features than those of the broad-faced grizzly.

John Marriott

Black bears begin hibernating in late October and November in the Central Rockies, falling into a deep slumber in a winter den that is usually a shallow hole beneath an uprooted tree or a mass of boulders. While true hibernating species experience a dramatic drop in body temperature and metabolic rate, black bears' systems never really slow down during their winter sleep, which makes it all the more amazing that they can go without food and water for five or six months.

In April, black bears begin emerging from their winter dens. Mating season occurs not long afterwards, and lucky people may find themselves spectators to the antics put on by a boar trying to impress a sow. Boars will hang suspended from tree branches, roll giant logs, and run up and down steep embankments, all in the spirit of the mating game. Eight months later, in the dead of winter, the sows

Photos: John Marriott

give birth to one to three tiny blind and hairless cubs. The young stay with their mother through the following winter until early in the second spring.

The Icefields Parkway and the Banff-Radium Highway are excellent spots to see and photograph black bears. Remain in your vehicle and observe all bears from a distance. Black bears that have become accustomed to humans often lose their natural fear of us and can be extremely aggressive and dangerous.

John Marriott

The front cover photo

Early each spring in the Central Rockies, the sun beats down and melts the snow at road edges. A few weeks or a month later, these open roadsides turn green and then sprout a wondrous spread of bright yellow dandelions, a favourite food of the black bear.

For the month of May and part of June, black bears frequent the roadsides enjoying their gourmet meals and munching away oblivious to their surroundings. The front cover photo was taken in June, 1996, in a field of dandelions on the side of the Trans-Canada Highway. I was able to photograph the bear portrayed and his smaller mate from my car for an hour without disturbing them before they ambled over the hill and out of sight late in the afternoon.

Gray Wolf

Canis lupus

Terry Willis

Driving from Banff to Canmore one day in January, 1996, I noticed three panicked elk nervously edging along the fence bordering the north side of the Trans-Canada Highway. Glancing up the hill to the west, I was startled to see seven wolves moving toward the elk.

A big gray wolf slowly pulled away from the others and began to run, pushing the elk into a fast-paced trot east along the fence. The wolves behind were joined by four black wolves from out of the trees above the highway, and together the ten of them fanned out along the hillside and began to lope after the lead wolf and the elk.

Within ten seconds, the elk took off running. The wolves followed, plunging through the deep snow at an obvious disadvantage to the long-legged elk. The race continued at roadside for almost a kilometre before the wolves and their prospective prey disappeared down a gully. Moments later, seven of the group of ten wolves reappeared and trotted back along the fence retracing their steps, then merged into the forest for the final time.

Terry Berezan

A week later I snowshoed along the old wolf tracks and found an elk carcass on the far side of the gully. The bones were bare, but wolf sign was all about: numerous sets of tracks larger than big German shepherd tracks, wiry hair-filled scat, and a central aspen stump where the pack had marked their possession, yellowing the snow with urine.

In winters of deep snow and long cold snaps, hardy and resourceful carnivores like wolves have banner seasons, feasting on the elk, deer, moose and bighorn sheep weakened by the harsh conditions.

Not surprisingly, it hasn't always been so easy for wolves in the Central Rockies Ecosystem. For much of the latter part of the nineteenth century and the start of this century, wolves were heavily persecuted as vermin and pests. They were absent from this area until the late 1960s, after being common throughout the Rockies prior to the arrival of explorers, trappers and settlers.

Gray Wolf

Canis lupus

John Marriott

Wolves are very social animals, living and hunting in territorial packs of three to 18 wolves. Each pack has a sophisticated social hierarchy, ruled by a mated pair of dominant wolves known as the alpha male and the alpha female. This pair usually mates for life, and each year produces four to eight pups in a secluded den.

Wolves are the largest of the wild dogs. An adult male can weigh up to 70 kg (150 lb), although weights around 50 kg (110 lb) are more normal. One of the easiest ways to distinguish between a wolf and a coyote is by size: wolves are about twice as big as coyotes.

One pack in Banff National Park has black, brown, gray and blond members, and this difference in colouration is another easy way to discern a wolf from a coyote. By comparison, a coyote is almost always gray with tawny or rusty streaks on its tail, legs, back and head. Other features that can help identify a wolf are its long legs and broad face (coyotes have shorter legs and narrow fox-like faces) and its straight tail (husky dogs always have curly tails that point upwards).

Wolves in the Central Rockies are extremely mobile, due in large part to their recent return to the area and the ongoing process of re-establishing territories. A pack may have a home range or territory as large as 3000 square kilometres, which is often passed on and defended from generation to generation. The howl of the wolf is used to communicate between individuals and packs in different locales, as well as to indicate the presence of a territory (scent-marking with urine serves the same purpose).

Wolves are usually extremely wary of humans, but will occasionally display tendencies of curiosity. I have twice had wolves approach within fifty metres of me before retreating. Other than humans, wolves have few natural enemies. In rare instances they have been known to defend their kills against grizzly bears or even take kills away from cougars.

The Central Rockies wolf population is fairly stable, but is under constant pressure from hunters and other human threats, including trains, vehicles and resource development. The call of the wolf has spurred governments in the United States into spending millions of dollars reintroducing wolves into areas from which they were extirpated, yet governments in Canada still sanction full-scale wolf kill programs and open hunting seasons.

Coyote
Canis latrans

Paul Smith

The yipping bark of the coyote is a sound often heard in the Canadian Rockies. Years ago, my friends and I were treated to the harmonious chorus of a family of coyotes in Kootenay National Park. To our delight, the yipping and barking reached a fevered pitch each time we let out our own soulful calls, and within minutes two coyotes ventured out of the trees below us and joined our yapping party.

Members of the dog family, coyotes are smaller than wolves but larger than foxes. An adult male coyote is about the same size as an Irish Setter or Golden Retriever, weighing up to 20 kg (44 lb). Females are usually a bit smaller. Both sexes have large bushy tails that are almost half the length of their bodies.

Coyotes are usually tawny greyish-brown in colour with red highlights on their tails, legs, backs and heads. They have facial features similar to those of a fox, with a long narrow nose and large pointy ears.

Wolves are twice the size of coyotes, and there is no love lost

John Marriott

between either species, though they do co-exist throughout much of the wolf's range. By comparison, coyotes and red foxes rarely co-exist, especially in areas like the Central Rockies where there is a healthy population of coyotes. Unlike wolves and most red foxes, coyotes are not overly wary of humans, and often become habituated to the presence of people.

Coyotes prey on small rodents and animals like deer mice, ground squirrels, and snowshoe hares. They will occasionally band together to hunt larger ungulates like deer and elk. Coyotes have perfected a technique called "mousing," where they patiently stalk a mouse or ground squirrel and then pounce high in the air and land on it. In winter, they use their excellent sense of hearing to listen for mice running beneath the snow, then pounce up and sharply downward, breaking the crust of the snow with their forefeet to catch their prey. In long hard winters, coyotes rely heavily on carrion; I have observed them feeding on leftovers from the kills of cougars and wolves several times.

John Marriott

Mated pairs of coyotes often stay together for several years, breeding each winter. By April or May, the female gives birth to four to eight pups in a hollowed-out underground den. For the first month or two, the female remains with the pups while the male goes out hunting and returns to the den to regurgitate food for the youngsters. By autumn, the pups are full grown and go out on their own, reaching sexual maturity the following winter.

In the late 1980s, a fence was strung up along both sides of the newly widened Trans-Canada Highway on the slopes of Mt. Norquay in Banff National Park to keep animals off the road and reduce bighorn sheep and deer mortality. Ingenious coyotes soon began using the fences to their advantage, running sheep up against the fence and trapping them. Wildlife officials soon recognized the problem and covered the fences with green plastic sheeting so the bighorn sheep could see the fence and escape without becoming trapped.

The fencing also caused a boom in the rodent population near the highway after officials decided to stop mowing the roadside grasses. The coyotes benefited immensely, but were liable to be hit by passing traffic. They face the same danger when they attempt to feed on roadkill, so stay alert while driving to avoid collisions with these wily creatures.

The resourceful coyote is well-adapted to the Central Rockies and is abundant throughout the montane valley bottoms in the mountain parks and foothill regions. While the enemies of coyotes include humans, wolves and mountain lions, they have persevered and flourished in the Canadian Rockies. Coyotes can often be spotted hunting in open meadows or at roadside at dawn and dusk.

Cougar
Felis concolor

John Marriott

The cougar or mountain lion is one of the more elusive animals in the Canadian Rockies. Common in the foothills and fairly common in more mountainous areas, these long, lithe cats are seldom seen because of their nocturnal and secretive nature, and their tendency to stay away from humans.

My only encounter to date with a cougar happened by chance in the spring of 1994. I was photographing two coyotes in an open meadow near the Trans-Canada Highway in Banff National Park, and couldn't figure out why they were ignoring me yet looking very apprehensive.

I glanced briefly to my left and stared straight into the eyes of a cougar hunkered down amidst tall grasses about ten paces from me. After an eternity of eyeing each other over, the cougar slowly backed away a few metres and lay down again, then closed its eyes and appeared to go to sleep.

Paul Smith

Although I couldn't see anything else at the time, it turned out that the cougar had killed an elk nearby and was guarding the carcass. The day after I had inadvertently disturbed the cougar from its kill, Parks Canada closed the area for several days until the cougar was finished feeding and had moved away.

Cougars are the largest wild cats found in the Central Rockies, with males reaching a total length of 2.5 m (8 ft) and weight of 70 kg (150 lb). Females are significantly smaller. Cougars usually sport a uniform coat of tawny brown. They are easy to distinguish from other cats such as the lynx because of their colour, size and thick long black-tipped tails. Both cougars and lynx have large feet and leave behind sizable clawless tracks that measure up to 10 cm across, particularly well-defined in snow.

Male cougars generally have larger home ranges than females, with territories as big as 900 square kilometres. Most males have a territory that overlaps with that of at least one female, although cougars are mainly solitary creatures, using urine and feces to mark the boundaries of their range.

Females can give birth at any time of the year to two to four spotted kittens, which remain with the mother for over a year learning how to hunt and survive on their own. Cougars prey on deer and elk, and will also take down moose, bighorn sheep and smaller animals like snowshoe hares. They are refined and specialised predators, using their stealth and cunning to stalk within pouncing range of their prey, then leaping forward in great bounds to overtake it. Cougars kill their prey by vaulting onto an animal's back and piercing the spine with razor-sharp canines, while simultaneously pulling back on the animal's head to break its neck.

Attacks on humans are extremely rare, particularly in the Central Rockies. Cougars have few natural enemies, but do face threats from hunting, habitat loss and human disturbance.

Terry Berezan

Driving on the Smith-Dorrien/Spray Road in Kananaskis Country one December morning after a snowfall, I noticed snowshoe hare tracks criss-crossing the road all over the place. Slowing to a near crawl, I picked up the tracks of a lynx amidst the hare tracks and followed them for a kilometre before they disappeared over a snowbank and into the trees.

Stopping my car, I stepped off into the snow to see where the tracks led, and promptly sank in up to my hips. Each step I took required a herculean effort, and the lynx imprints that seemed to float along on top of the snow beside my great depressions didn't help matters.

Uniquely designed to revolve around the life of the snowshoe hare, their primary prey, lynx have giant fur-covered feet that make their tracks even a little larger than those of their big cousin, the cougar. The feet of lynx act as snowshoes, allowing them to move over snow at great speeds and catch the fleet-footed hares.

Lynx
Felis lynx

Paul Smith

The lynx I followed in Kananaskis Country had obviously moved off the road with a purpose, because a few steps from the bank I spotted a pair of snowshoe hare feet sticking up in the air, attached to nothing but bloody snow and hair.

Lynx populations are intricately tied to the cyclical snowshoe hare populations in the Central Rockies, so every eight to ten years when the snowshoe hare population crashes, the lynx population soon follows suit. However, lynx do prey on other animals, such as mice, squirrels, and the occasional marten.

Lynx frequent areas that support snowshoe hares, including valley bottoms and young coniferous forests. They tend to be more common near Lake Louise and north toward Jasper than in the southern part of the Central Rockies.

An adult lynx weighs approximately 10 kg (22 lb) and has a tawny grey mottled coat with a short black-tipped tail and long black ear tufts. Mating occurs in early spring, and two months later one to five sightless young are born in a den among boulders or under an outcropping. The kittens stay with their mother for one winter and branch off to lead their own lives the following spring.

Lynx have few natural enemies, although wolves and cougars will occasionally kill them. The more hapless kittens are preyed upon by owls, coyotes and foxes.

Since they are primarily nocturnal, lynx are rarely sighted and their tracks are often the only reminder that they're about. They do not often overlap ranges with bobcats, although where they do, lynx can be distinguished by their long legs, big feet, facial ruff and longer black ear tufts. In contrast, bobcats often resemble large housecats in appearance.

Wolverine

Gulo gulo

Pat Morrow

Few animals garner more respect than the wolverine, which has achieved mythical status in mountain folk tales as a beast of unparalleled strength and cunning for its size. In ancient aboriginal folklore, the wolverine was a trickster, but was also one of the original and most important Animal People, with great powers of healing and transformation.

Half-predator and half-scavenger, the wolverine is a critical indicator species in the Central Rockies Ecosystem. In other words, because wolverines depend on vast wild areas that feature a diverse predator base including wolves, and a diverse prey selection for carcasses, such as elk and moose, the wolverine population is an indicator of how well the animals in the area are doing. If wolves begin to disappear and there are fewer carcasses available to scavenge, then the wolverine population immediately starts to decline, reflecting the loss of diversity caused by the drop in wolf numbers.

There is no estimate of how many wolverines there are in the Central Rockies, and sightings are rare. Solitary and elusive creatures, wolverines often have home ranges of up to 2500 sq km and can cover 60 km in a day, traversing the most inhospitable terrain.

Bear-cub sized members of the weasel family, an adult male wolverine usually weighs about 15 kg (33 lb) and can be up to 1 m (3 ft) long from nose to tail. Wolverines have rich dark brown fur and small heads. They are often called "skunk-bears" because of two pale brown stripes that run from their shoulders to the base of their tails. The nickname also refers to their habit of marking their food with a nasty-smelling musk.

Wolverine fur is a valuable commodity because of its luxurious nature and its frost-free characteristics. Many wolverine populations outside of the Central Rockies, particularly those in northern Canada and Alaska, are heavily trapped because of this precious fur.

Like many other members of the weasel family, wolverines breed in the spring but the females don't actually give birth until the following spring so that the young can have a full summer to grow and increase their strength for the upcoming winter. Litters usually range from one to five young, but do not necessarily occur each spring, particularly if the preceding winter was a harsh one.

Wolverines have an excellent sense of smell, allowing them to sniff out carcasses from great distances. They come equipped with powerful jaws and large sharp teeth, which are particularly useful for crushing bone and frozen carrion.

While I have yet to see a wolverine, former Banff National Park interpreter Jeff Waugh did when he was camped near a high alpine pass one summer. Jeff and his wife Candace had spent the previous day watching a grizzly and its cub move down the slope opposite them. That particular morning, Jeff unzipped the tent flap just in time to catch a wolverine passing by fifty metres from the tent on its way up through the pass.

Marten

Martes americana

Paul Smith

For a small carnivore, a marten packs a lot of punch. Just 60 cm (24 in) long and weighing less than a kilogram (2.2 lb), martens are feisty and temperamental creatures. They are not afraid to display their displeasure with the occasional human visitor in their fiercely guarded territories, although they will often approach a person out of curiosity.

Common throughout mature coniferous forests in the Central Rockies, the marten is also called the pine marten or American sable. Martens are characteristic members of the weasel family, and are about the same size as mink, but smaller than their cousin and enemy, the fisher. Martens count great horned owls, lynx and wolves among their other enemies.

Like many of its weasel relatives, the marten is territorial, scent-marking its home range with its anal musk glands. Females often have smaller territories than males, with a higher concentration of food to help support themselves and their young. Martens mate in late summer, which is the only time they lose their cantakerous nature. Nine months later, the females give birth to two to three young.

Martens are extremely agile and quick both on land and in the trees, and use this natural ability to chase down red squirrels, grouse, voles, mice and other small rodents. They will occasionally pull down larger animals such as snowshoe hares given the right opportunity, and are not above dining on carrion.

One summer in Mount Assiniboine Provincial Park I watched a marten engage in a furious battle with a Columbian ground squirrel that was at least half the marten's size. Not keen on being a marten's lunch, the ground squirrel finally broke free after raking its claws along the belly of the marten. The loss and the injury did not prevent the marten from turning and racing towards me in a bluff charge before climbing the nearest tree and hissing at me from a safe height.

A marten looks a lot like a ferret, although larger, with a lustrous chocolate-coloured coat and a 15 cm (6 in) long bushy tail. All martens have distinctive yellowish-brown markings on their throat and chest, with a light brown face.

Mink

Mustela vison

Doug Leighton

Mink are similar in size to martens, but are stockier and darker brown in colour with a characteristic white chin and occasional white markings on the chest and belly. A member of the weasel family, an adult male mink can weigh up to 1.5 kg (3.3 lb) and measure 65 cm (25 in) in length. Females are usually quite a bit lighter and smaller.

Superb swimmers, mink are able to make use of both land and water. They can be found in many Central Rockies wetlands and along creeks and streams. Banff author and naturalist Mike Potter observed a mink slipping in and out of the open water repeatedly from the edge of the ice in Stewart Canyon in Banff National Park one winter. Potter believes the mink was diving in search of one of its favourite foods, trout. Besides fish, mink feed on small birds, voles, muskrats, and frogs, as well as a host of smaller aquatic and terrestrial creatures. They have also been known to stock extra food for lean times that may lie ahead.

Mink den in creekside burrows that are often the abandoned lodges of beavers or muskrats. Like other weasels, mink use scent glands to mark their territories. They are extremely defensive of their home range when other mink trespass.

Litters range from two to six kits born in the spring. By autumn, the young are out on their own establishing a territory.

Mink are fairly common and primarily nocturnal. They are occasionally sighted at dawn or dusk in close proximity to water.

Al Williams *Least weasel*

There are three species of smaller weasels found in the Central Rockies: the long-tailed weasel, the short-tailed weasel and the least weasel. These three weasels share common characteristics, including long, narrow tube-like bodies with short legs and compact ears, and distinct summer and winter pelages.

Voracious killers for their size, these weasels feed primarily on small rodents like mice and voles, but will also take down larger animals such as ground squirrels, red squirrels and snowshoe hares. They kill their prey by piercing the neck vertebrae or the brain of the animal with their long, sharp canine teeth.

Like other relatives in the weasel family, breeding occurs in the summer. However, the four to eight young are not born until the following spring due to delayed implantation of the embryos. Enemies of these smaller weasels include owls, hawks, coyotes and larger members of the weasel family.

Long-tailed weasel *Mustela frenata*

Long-tailed weasels are distinguished from the two other smaller weasels by their size and the length of their tails. An adult male can be up to 45 cm (18 in) long, including its 15 cm (6 in) long tail which is twice the length of a short-tailed weasel's. Males weigh 150-200 g (6-7 oz), while females are about twenty-five percent lighter and shorter.

Long-tailed weasels have black-tipped tails that stand out in the winter, although the rest of their all-white coat provides excellent camouflage against the snow. Their summer coat is a lustrous brown on top and pale yellow below.

Long-tailed weasels prey on larger animals than the two smaller species, with birds making up a larger percentage of their diet.

The long-tailed weasel is more common in the prairies and foothills than it is in the mountains, so although it is active during the day and night, it is rarely seen in the Central Rockies.

Short-tailed Weasel *Mustela erminea*

Short-tailed weasels (also called ermines) are larger than least weasels but smaller than long-tailed weasels. Males are 25-30 cm (10-12 in) long, including black-tipped tails 5 to 8 cm long, and weigh 80 g (3 oz). Females are approximately two-thirds the length and weight of the males.

Their fur is all white in winter except for the black tip of the tail, while in the summer it is brown on top and buff yellow beneath.

Short-tailed weasels are abundant in the subalpine zone, preferring its combination of meadows and forests. However, despite the fact that they are the most common of the three smaller weasels in the Central Rockies, short-tailed weasels are rarely seen because of their nocturnal nature.

Least Weasel *Mustela nivalis*

The least weasel, as the name suggests, is the smallest of these three species. In fact, it is the smallest of the North American carnivores, tipping the scales at a mere 50 g (2 oz) - about the same weight as a least chipmunk - and measuring just 20 cm (8 in) long.

Least weasels do not have black tips on their short tails like short-tailed or long-tailed weasels, so their white winter coats blend perfectly into the snowy background. In early spring, least weasels molt and their coats turn a rich dark brown on top for the upcoming summer, while staying white underneath.

Voracious feeders, least weasels often consume half their weight in rodents each day. They are active only at night and are distributed irregularly through meadows and forests in the Central Rockies.

Moose

Alces alces

John Marriott

Moose are the largest members of the deer family and the largest land animal naturally occurring in the Central Rockies (bison are larger, but have been absent from the Rockies for over a century). Adult bull moose can reach 3 m (10 ft) in length and 2 m (6.5 ft) in height at the shoulder, with weights in excess of 500 kg (1100 lbs). Females, called cows, are typically one-tenth smaller.

An adult moose looks a bit like a long-legged, dark brown horse with a large long nose and a prominent shoulder hump, but without the long tail or mane. For much of the year, it is easy to tell bulls and cows apart because the bulls have antlers and the cows do not. It is more difficult in the middle of winter after the bulls have dropped their antlers; I once crept within twenty metres of four moose, trying to determine how many bulls there were, and could

Mike Potter

only tell for sure once I saw the antler stubs up close.

Like elk and deer, moose antlers drop off and regrow each year. However, unlike elk and deer, they don't fork from a main beam. Instead, moose antlers are palm-shaped with spikes pointing off the front of the broad surfaces (like an open hand facing the sky with the fingers pointing out from the palm). Each antler can weigh as much as 35 kg (75 lb). The antlers that are cast off each winter are gnawed by rodents such as deer mice and porcupines and are important sources of nutrients for them.

While appearing gangly at times, moose are actually agile animals for their size, and can run swiftly through dense forest or deep snow. They are excellent swimmers and can dive to depths of 5 m (16 ft) to feed on bottom-dwelling plants in ponds and lakes.

Moose
Alces alces

John Marriott

Moose are fairly common throughout the Central Rockies, although they have struggled in recent years in the Bow Valley, the core of the Central Rockies Ecosystem. Parasitical liver flukes, collisions with trains and highway traffic, wolf predation, and increased competition for forage from elk have led to this decline.

In winter, moose are found in mixed forests feeding on the twigs of willows, aspen and red osier dogwood, while in summer, they frequent low-lying marshlands in search of water plants. The fall is a gruelling time of year for the bulls if they are big enough to take part in the rut. The bulls with the largest antlers usually acquire the rights to mate with the cows, although when bulls with like-sized antlers meet up there is often a vicious battle to determine supremacy. Dominant males often expend much of their fat reserve during the

John Marriott

rut, which weakens them for the upcoming winter and makes them more susceptible to predation from wolves or cougars.

In the spring, cow moose can give birth to one or two calves weighing between 10 and 15 kg (22-33 lb) each. The calves do not have protective camouflage patterns like elk calves and deer fawns, relying instead on an aggressive and protective mother. Cow moose with newborn calves, or bulls during the rut, are extremely dangerous to approach and will attack humans if provoked.

Moose are particularly fond of salt, and can be seen at a number of different natural roadside mineral licks in the Central Rockies, including excellent ones along Highway 93S in Kootenay National Park and near the Natural Bridge in Yoho National Park.

Elk

Cervus elaphus

John Marriott

Few moments in the wilds of the Rockies are quite as exhilarating as hearing the trumpet-like bugle of a bull elk on a crisp September morning. It is a call to challenge from one majestic male to another, ringing through the air during the spectacular fall rut.

Visitors and locals alike can observe the spectacle as the biggest and strongest bulls vie for the right to plant their genes with as many females as possible. The males herd the cow elk into groups, called harems, to keep track of the cows and discourage them from running away, while at the same time staving off the many would-be rival suitors.

When two bulls of equal strength face each other, the battle can be tremendous. The bulls rush together and lock antlers, pushing and tugging to try to gain an advantage or gore the other bull. While the battles rarely lead to death, the rut does take its toll on the bulls, many succumbing that same winter to the cold or to predators like the wolf.

I followed a big bull I had nicknamed "Moose" (for the moose-like palmated antler he had on one side) through the rut one fall,

John Marriott

watching him herd thirty-five cows together and fight off challenges from at least two other big bulls. The energy he expended weakened him considerably, making the long winter that much more difficult for him. By the following spring, "Moose" had a bloated gut and looked like he might not make it to the summer. Indeed, three weeks later, just after dropping his antlers, he retreated to a dense grove of trees and passed away.

In late May or June each spring, females that were impregnated in the autumn and survived the winter in good health give birth to one calf (occasionally two). The helpless newborns are spotted white on their light brown coats as a camouflage against predators, which include grizzly bears, wolves, cougars and coyotes. Their protective mothers become extremely aggressive towards humans and dogs, and will attack if threatened, using their hooves and teeth as weapons.

Elk, or wapiti, are the most common large mammal in the Central Rockies. Given the aggressive behaviour of the cows in spring and the bulls in fall, it's easy to see why elk, not bears, are the most dangerous animal in these mountains as far as humans are concerned.

Elk

Cervus elaphus

John Marriott

Elk are larger and a darker shade of brown than deer, but smaller and lighter in colour than moose. Elk are about 1.5 m (5 ft) high at the shoulder, with big bulls weighing as much as 400 kg (880 lb) and cows weighing about 250 kg (550 lb). All elk have light-coloured rump patches, short tails, and dark brown manes on their necks, although a bull's mane is generally more pronounced.

Elk are more social and gregarious than moose, often forming herds that are segregated by sex and even age. They are extrememly adaptable and as many as one thousand elk live in or near the town of Banff in Banff National Park. By doing so, they take advantage of the lush gardens and lawns in the town and avoid predation by wolves or cougars that are too wary of humans to venture near the town's limits.

Elk prefer the montane valley bottoms for much of the year, feeding on grasses, deciduous shrubs, and even the white bark of aspen trees (in aspen stands throughout the Rockies the dark black girdled base of each tree is evidence of elk chewing on it in the past). However, particularly in the summer, one can see elk just about anywhere in the Central Rockies. I have twice seen elk at elevations above 2300 metres in the alpine.

For most of the year, bulls are easy to distinguish from cows because of their antlers. From early spring until late summer, the antlers grow continuously, protected by a velvet covering. Just before the rut, the bulls shed the velvet and the antlers calcify into hard bone. The largest, healthiest bulls usually grow the biggest antlers, and it is these same bulls that dominate the rut and get to breed. The bulls' antlers, which can weigh up to 20 kg (44 lb) each, fall off in March every year.

The area surrounding the town of Banff in Banff National Park is prime elk habitat and places such as Vermilion Lakes and the Banff Springs Golf Course are ideal for spotting the big ungulates.

John Marriott

Woodland Caribou

Rangifer tarandus

Mike Potter

Woodland caribou maintain a precarious hold in the Central Rockies Ecosystem, at the south end of their range. In the Central Rockies, woodland caribou move down to the valley bottoms for the rich spring growth, then follow the greenery up to the alpine meadows for most of the summer. In late fall, with the first deep snowfalls, they return to the forested valley bottoms and remain there until the snow deepens enough that they can reach the lichens hanging from old spruce and fir trees at higher elevations. Caribou are equipped with unusually large hooved feet that allow them to move about in the deep snow without sinking.

The woodland caribou is a member of the deer family and is slightly smaller in size than an elk. Most caribou are brown to dark brown in colour, with off-white neck manes and brown heads. Adult bulls tip the scales at about 200 kg (440 lb), while cows weigh about 140 kg (310 lb). Unique to our deer, both bulls and cows carry antlers, growing and dropping them at different times of the year. The bulls' antlers are identical to those of the northern caribou featured on the Canadian twenty-five cent coin, while the cows' antlers are smaller, less intricate versions.

John Marriott

On a backpacking trip through the northern reaches of the Central Rockies one July, I happened upon a cow caribou suckling her tiny calf high in an alpine meadow. The cow spotted me from 300 metres away and the two of them immediately raced off across the meadow and out of sight over a nearby ridge. Unlike moose and deer, which often have twins, woodland caribou generally give birth to one small brown calf a year near the end of spring. Like most ungulate young, caribou calves are quickly on their feet just hours after being born and can outrun a human within a week. Woodland caribou are skittish animals, particularly when they have young, but they can also display an uncanny curiosity at times.

Woodland caribou rarely move about in groups of more than ten. The only caribou found in Banff National Park are two dozen or so in the remote northeastern sector of the park. Human use of the area these caribou frequent is now restricted and the area has been designated a special preservation zone for their benefit. There are also small populations of woodland caribou in the Alberta wilderness areas bordering Banff National Park at the top end of the Central Rockies Ecosystem.

Mule Deer

Odocoileus hemionus

John Marriott

Mule deer are common throughout western North America, and are more plentiful in the Central Rockies than the white-tailed deer. Mule deer get their name from their large wide mule-like ears. Smaller than elk, caribou and moose, adult males (bucks) weigh on average 115 kg (250 lb), though they can attain weights in excess of 200 kg (440 lbs). The females (does) are considerably smaller.

Even for people who have seen a lot of deer, it can be difficult to tell a mule deer apart from a whitetail. I have spent the last few years repeatedly testing my wife by spotting a deer off in the distance and asking her if it is a mule or a whitetail, teaching her the easy ways and the hard ways to decipher between the two. In the Central Rockies it can be handy to know the differences because the two species are often seen in close proximity to each other, particularly on the Bow Valley Parkway in Banff National Park and on Highway 40 in Kananaskis Country.

John Marriott

A mule deer has a narrow cylindrical white tail with a black tip, seen against a white rump. When mule deer run, the tail remains pointed down. In contrast, white-tailed deer have a wide brown tail with a white underside that is only visible when the deer becomes alarmed and raises its tail like a white warning flag.

Mule deer bucks begin to grow their antlers each year in the spring and drop them after the rut, in December or January. Their antlers branch out in a Y-shape from the main beam; depending on how old and healthy the buck is, the antlers will branch again in a Y-shape from each of the tines on the first Y (white-tailed bucks' antlers have tines that all grow off one main beam).

The rut takes place in November and early December, during which the necks of the big bucks become swollen, indicating that they are ready to breed. The dominant bucks are polygamous breeders, often moving from one female to the next in short succession.

John Marriott

Does have one to three fawns each June, immediately licking the newborn clean to eliminate odours that might attract coyotes or other predators. For the first few months, the fawns are chestnut-coloured with white spots on their coat to camouflage them against their surroundings. I have seen fawns literally melt into the background and disappear from sight even when I knew they were still there in front of me. One summer in Kananaskis Country I stumbled upon two young fawns lying in a meadow. After I had turned my head to look around for their mother, I returned my attention to the fawns and could not pick out one of them against the tall brown grasses. It was only when the fawn moved its head that I saw it again, still lying in exactly the same spot!

Adult mule deer are yellowish-brown in summer with a hint of grey, but have pale yellowish-white undersides. In winter their coat is more grey than brown and is more uniform overall with longer hair. Throughout the year, mule deer have a distinctive white patch beneath the chin to the front of the neck.

Mule deer are both grazers and browsers, eating a wide variety of shrubbery, plants and grasses, including the twigs of aspen and

John Marriott

Douglas-fir in winter. They are a favourite prey species of cougars, but also suffer predation from wolves, bears and the occasional coyote or lynx.

When a mule deer gets excited and bounds away from a person or a predator, it is an interesting sight. Mule deer run by jumping up and forward with stiff legs as if they are on a pogo stick, called stotting. By comparison, white-tailed deer run in great stretching forward leaps.

Several years ago, mule deer frequented the parking lot of the Upper Hot Springs in Banff and became used to taking handouts from visitors. When the deer started to get aggressive towards people, slashing with their front hooves, Parks Canada personnel had to put down many of them. Today, visitors are made well aware that it is illegal to feed wildlife, so that situations like that at the Upper Hot Springs can be avoided.

Mule deer are easiest to spot at dawn and at dusk, and may be observed in many parts of Kananaskis Country or along the Bow Valley Parkway and the Mount Norquay Road in Banff National Park.

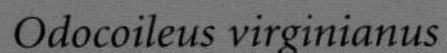

White-tailed Deer

Odocoileus virginianus

John Marriott

On my way back along a trail after photographing the autumn colours by the Kananaskis River one October, I suddenly noticed an enormous white-tailed deer buck off to my left. I had just enough time to count twelve or thirteen points on its impressive rack of antlers before the big buck, which must have weighed close to 130 kg (280 lb), let out a snort and dashed away. It ran flat-out across the meadow, leaping in great arcs until it disappeared into the trees three hundred metres away.

It was one of the first times I'd seen a white-tailed deer run, leaving me amazed at how different and graceful it looked compared to the many mule deer I'd seen bounding away from me in stiff-legged jumps. The whitetail seemed to glide across the meadow, eating up vast distances with each leap, in sharp contrast to the mule deer that always look like they jump up as much as forward.

White-tailed deer are common in certain parts of the Central Rockies, including much of Kananaskis Country, the eastern part of

John Marriott

Banff National Park and the southern regions of Kootenay National Park. They prefer the montane valley bottoms, but can also be found in the subalpine and alpine during the summer. Typical white-tailed deer habitat includes young forest stands and open woods interspersed with meadows and clearings.

White-tailed deer can be distinguished from their close relatives, the mule deer, by their tails and their antlers, as well as by a number of more discrete features, such as their differences in running style. White-tailed deer, as their name implies, have tails with a white underside that flip up as a visual signal to other whitetails that danger may be present. Although the outer surface of the tail (seen when a whitetail is unalarmed) is brown with a white edge, the white underside is easily visible whenever the deer erects its tail like a warning flag. By comparison, the tail of a mule deer–which is narrow, white and has a black tip–always points down.

White-tailed Deer

Odocoileus virginianus

Terry Berezan

White-tailed bucks have antlers that grow from one main beam, with each separate tine coming from that beam. Mule deer have a main beam that forks into two separate branches and then may branch once or twice more. White-tailed does do not have antlers.

White-tailed deer are extremely adaptable; like the coyote, they seem able to thrive almost anywhere regardless of conditions. They are the most widespread deer in North America.

Whitetail does usually produce one or two fawns in the spring. Similar to mule deer, the fawns are scentless and are rusty brown or chestnut coloured with dappled white spots on their coats to help them blend into the vegetation. Fawns are at risk from predation by wolves, bears, coyotes, lynx, cougars and larger members of the weasel family, while adults only have to worry about wolves, bears and cougars for the most part.

John Marriott

Adult white-tailed deer have two distinct coats. They are reddish-brown in the summer, and greyish-brown in the winter. Besides the white underside of the tail, white-tailed deer are white on the belly, as well as on the rump and just below the chin. They browse more than they graze, preferring to eat twigs, buds and leaves of aspen, red osier dogwood and other deciduous shrubs and trees.

Whitetails, which are generally smaller than mule deer, can be seen along Highway 40 in Kananaskis Country, the Bow Valley Parkway in Banff National Park and Highway 93S in Kootenay National Park.

Mountain Goat

Oreamnos americanus

John Marriott

The mountain goat probably didn't have a chance. A powerful surge of snow sweeping down the slope from the cliffs above caught her in its fury and deposited her body at the bottom, just ten feet from the trail.

I happened upon the glittering bones a month later, scavenged bare by a grizzly and a host of other animals and birds. I found the skull up the hill a bit, still intact and nestled between bright red paintbrushes and golden glacier lilies.

Looking around, I noticed I was in the middle of perfect goat habitat. The rich green slopes above me were bordered by steep cliff bands that stretched across the skyline, providing excellent browse in close proximity to the safety the cliffs afforded; predators did not dare attempt the rocky faces and crevices for fear of falling.

From birth, the jagged cliffs and mountain tops that deter all but the hardiest and most nimble of creatures are home to the thousands of mountain goats that inhabit the Central Rockies. Equipped with suction-like cups on the bottoms of their hooves and an amazing dexterity that includes being able to bend their bodies around backwards to change direction on narrow exposed ledges, mountain goats

John Marriott

are actually more comfortable on cliffs and rocky terrain than they are on flat land.

When they are on level terrain, the mountain goats' enemies include cougars, wolves and grizzly bears. Grizzly bear researchers recently discovered that the Odaray Plateau in the Lake O'Hara region of Yoho National Park plays an important role in grizzly bear predation on mountain goats. The goats are occasionally caught on the open flat terrace by the big bruins, so to eliminate the potential for human conflicts and disturbances, Parks Canada has permanently closed the area to summertime human use.

On the rocky cliffs and crags that mountain goats prefer, their chief predator is Mother Nature. Most goats lose their lives to avalanches, rock slides or misplaced steps, although young goats (called kids) are occasionally knocked from their airy domain by golden eagles.

Goats are often mistaken for bighorn sheep, but are easily identified by their all-white (sometimes yellowish) coat, leggings and beard, and their short, black dagger-like horns. It is difficult to tell a male goat from a female, since both sexes have similar horns.

John Marriott

However, an adult billy goat is about twenty-five percent larger than a nanny, weighing up to 85 kg (190 lb), measuring 2 m (6 ft) in length, and standing 1 m (3 ft) tall at the shoulder.

Mountain goat courtship begins in late fall and extends into early winter. The billies soil themselves with their own urine and use this 'perfume' to attract the nannies. Half a year later, in June, the nannies retreat to a secluded ledge and give birth to one tiny kid (occasionally two). The young, which can follow their mothers around within a day or two, quickly learn the ropes of living in steep terrain.

Each spring, when the sun begins to beat down on the Rocky Mountains, mountain goats begin to molt, shedding their shaggy long-haired winter coats in favour of cooler short-haired coats. During this process, the goats often look unkempt and scraggly, with big chunks of hair falling off at a time. The warm winter coat begins to grow back in October or November, just in time for the goats to

Terry Berezan

retreat up the mountainsides to higher elevations in preparation for the upcoming winter (while most ungulates move down to the valley bottoms during winter, mountain goats seek out the tiny plants and grasses found on windswept snowless ridges and ledges).

Grasses and forbs make up the majority of a mountain goat's diet, but they will also eat the needles of coniferous trees to sustain them through long winters. Mountain goats get a limited supply of nutrients and minerals with their forage, so they will often travel kilometres from their normal haunts to the valley bottoms to reach mineral salt licks and satisfy their cravings.

Mountain goats are not true goats at all, being more closely related to the mountain antelopes of Europe and Asia. They can often be spotted along the Icefields Parkway in Banff National Park and at natural mineral licks along Highway 93S in Kootenay National Park.

Bighorn Sheep

Ovis canadensis

John Marriott

A seasoned hiker I know was hiking in Banff National Park one fall when he heard a thunderous gunshot erupt from the ridge above him. Not knowing what to expect or see, he crept up the hillside slowly, hearing three more terrifying crashes before he reached the ridge top and peered over apprehensively.

To his surprise, there were no people in sight, but instead a couple of bighorn rams. As he watched, the two rams suddenly reared up on their hind legs and rushed together, culminating in a resounding crash of horns as they clashed. The boom echoed off the cliff to the right and came back sounding exactly like a gunshot.

It is said there are few spectacles in the wild as rousing as that of a fight between two bighorn rams, a battle of the titans. One November I watched five rams smash each other about for an hour on the golf course in Radium Hot Springs before an embattled mean-spirited ram, with horns that curled past his eyes, arrived on the

John Marriott

scene and dispersed the five would-be challengers to his crown.

The heavy brown horns that adorn the heads of bighorn rams give the species its common name. Female bighorn sheep, called ewes, also sport horns, but much smaller, less-curled versions. Lambs, ewes and rams all have white rumps, short brown tails, white noses and tan-brown pelages, although lambs are often lighter in colour than the adults. Bighorn sheep have short, stocky builds, with large rams tipping the scales at 125 kg (275 lb), standing 1 m (3 ft) tall at the shoulder and measuring 1.7 m (5.5 ft) long. The ewes are considerably smaller and lighter.

Unlike the antlers of members of the deer family, the horns of bighorn sheep and mountain goats don't fall off each year. The horns are comprised of keratin, the same substance in fingernails. Growth slows to a virtual stop each winter, but resumes in the spring and summer, creating rings like those of trees.

Bighorn Sheep
Ovis canadensis

Mike Potter

Mountainous terrain, with open slopes and meadows for grazing, is ideal bighorn sheep habitat. They are not good at digging through deep snow, so they prefer areas that do not have high annual snowfalls, such as the foothills in Kananaskis Country. Almost as accomplished in climbing as mountain goats, bighorns use rocky crags and cliffs to escape predators. They spend much of their time eating grasses, which make up about 60 percent of their diet. Bighorns also eat sedges and wildlflowers, only browsing on shrubbery when food is scarce.

Ewes are impregnated during the November-December rut, and give birth in the spring to one tiny lamb (occasionally two). Like other ungulate young, the lambs are quickly on their feet and dashing about the hillsides. For most of the year, the gregarious bighorns form social herds, including summer nursing groups comprised of ewes and their young seeking protection in numbers. I ran into a gigantic herd of 35 lambs and 47 ewes high in the alpine in the White Goat Wilderness Area in the north end of the Central Rockies one summer.

John Marriott

Bighorn ewes are often mistaken for mountain goats, but are a different colour than the white or off-white goats and have rough brown horns as opposed to goats' sharp daggerlike black horns.

Adult and young sheep are preyed upon by cougars and wolves, while lambs are occasionally also taken by bears, lynx and coyotes. Bighorns are quick and agile runners, equipped with an above average sense of smell and excellent eyesight to help them avoid predation. Studies indicate that bighorn sheep may be able to see as well as humans who are using binoculars.

Bighorn sheep are plentiful in the Central Rockies, and can be viewed at high and low elevations throughout the area. Watch for them on roads near the Banff townsite, on Highway 40 in Kananaskis Country and on Highway 93S in Kootenay National Park.

Beaver

Castor canadensis

Terry Berezan

One morning on a drive through Kananaskis Country, I came upon a beaver slowly dragging a thick aspen tree across the road. I pulled off and followed the beaver into a nearby gully, where I discovered a beaver family, including four half-grown kits, hard at work moving trees and branches.

The focal point of activity was a dam on a tiny brook at one end of the gully. The dam had created a large, deep pool of water that stretched across the bottom of the gully, providing a safe travel route for the family back and forth between their lodge and the aspens. The lodge was built from the same materials as the dam: sticks, logs, and rocks pasted together with mud and dirt. Beaver lodges feature a snug inner chamber with underwater entrances leading to the deep parts of the pool (river-dwelling beavers often use burrows in the muddy banks as an alternate to constructing a lodge).

A beaver's lodge serves as a social gathering place, a shelter during winter, and a safe retreat should a predator happen along. Beavers are excellent swimmers, but are not particularly quick or agile on land, making them fair game for wolves, lynx, bears, cougars and coyotes on land. Their chief predator in the water or in their lodges is the river otter.

While they do not hibernate, beavers are much more visible during the warmer months of the year. In winter, they rarely venture far from their lodge surroundings, feeding on branches cached during the fall in the mud at the bottom of their pool.

Beavers are fairly common in marshy areas and wetlands throughout the Central Rockies. They are the largest rodents on this continent, with adults averaging just over 1 m (3 ft) in total length

Terry Berezan

and 20 kg (44 lb) in weight. Beavers are dark brown animals, with small blunt heads on chunky round bodies. A unique feature is a wide, flat hairless tail that looks scaly. The tail is used as a rudder in the water, and helps prop them up when they stand to chew on a tree. If a beaver is disturbed or threatened while it is in the water, it will smack its tail sharply on the water's surface as a warning to other beavers just before it dives to safety.

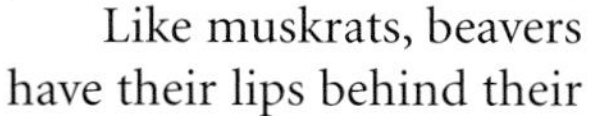

Like muskrats, beavers have their lips behind their long, sharp incisors so they can chew underwater. Other adaptations for their unique lifestyles include clear membranes that cover and protect their eyes while they swim, and self-plugging ears and nostrils that allow them to dive without hesitation. Beavers are also equipped with large, webbed back feet that help propel them in the water, and small, hand-like front feet which they use to grip and carry small objects such as branches or stones.

Beavers eat the buds, leaves and twigs of deciduous trees such as willows, poplars and aspens, but love the chewy cambium layer of the bark in particular. They also feed on aquatic vegetation including grasses and sedges. One adult can require up to 200 trees each year to subsist. Using its incisors, which resemble big buck teeth, a beaver can chew through a tree in a few minutes. Once the tree is down, the beavers feed on it and/or use it as part of their lodge or dam.

In beaver society, the females are at the top of the social ladder. Each beaver community consists of a mated pair, this year's offspring (1-8 kits born in late spring), and last year's offspring. Once the juveniles are full-grown and in their third spring, they are booted out of residence to start up their own community in a different location.

Beavers are active from dusk until dawn.

Porcupine
Erethizon dorsatum

John Marriott

In the summer of 1994, I was awakened in the middle of the night by something scratching at my tent while I was camping in the Upper Spray Valley in Banff National Park. My first thoughts were that the creature outside was a bear, and that I was in big trouble.

After a few minutes of heart-pounding terror, I got up the nerve to grab my flashlight and venture into the night to see what was out there. As soon as I unzipped the tent flap, I could hear an animal start to shuffle away, and in the beam of my flashlight I caught a porcupine waddling slowly towards the bush.

The next day I got the opportunity to watch this fascinating rodent as it gnawed on the trees all around my tent, and attempted once or twice to chew on anything of mine that had a salty residue from my touch. Porcupines eat green leaves in the summer and the sweet cambium (the layer just below the bark) on coniferous trees in the winter - they do not eat the bark itself. A porcupine can kill a tree by girdling it (stripping the bark off), but this in turn allows other trees to become established. Porcupines will rarely pass up a chance to supplement their mineral intake by nibbling on salty objects, so store your belongings securely if you head into the backcountry.

Porcupines waddle around quite slowly in general, but can move quickly when alarmed. They are excellent climbers and often scale trees to escape danger. Porcupines are protected by some 30,000 sharp, spine-like barbed quills that cover their body except for the front of the face, the short legs and the underbelly. When attacked, a porcupine will try to protect its face between two rocks and will swing its large quill-covered tail back and forth in the direction of the predator's attack - they don't actually throw the quills. If an animal is hit by the tail, the quills detach and lodge into the attacker, incapacitating it or even killing it. The fisher, one of the few carnivores that has devised a system to get by the dangerous quills, will slash at the vulnerable nose or will attempt to flip the porcupine over to expose its belly.

Porcupine quills are creamy yellow with black tips. The long guard hairs are yellow, while the face and feet are usually black. Porcupines are the second largest rodent in the Central Rockies (only beavers are larger), weighing up to 8 kg (17.5 lb).

Porcupines engage in an extraordinary and arduous mating ritual each fall that includes a lot of moaning, screaming, grunting and purring. The nose-rubbing affairs get serious when the male soaks the female in urine, at which point she lifts her tail to allow the male to mate with her.

Female porcupines give birth to one large baby each spring. Surprisingly, the young are not born blind and naked like most rodent babies. In fact, the newborn are covered in soft quills and fur, and can see and move about immediately. The young stay with their mother for a year and then embark on their own in the second spring when they reach sexual maturity.

Porcupines are common in the Central Rockies, particularly in coniferous subalpine forests. Watch for them in areas such as Lake Louise and Lake O'Hara in Banff and Yoho national parks.

Hoary Marmot

Marmota caligata

John Marriott

The shrill piercing call of the hoary marmot, or "whistler," is a common greeting for a hiker or backpacker going through open rocky alpine meadows and slopes of the Central Rockies. Marmot sentries, assigned to watch the ground and sky for predators, issue their warning whenever a human approaches a marmot colony.

After the initial alert, the marmots' curiosity generally takes over, and many of them will get up on perches near their burrow entrances to look over the intruder. I have camped near marmot colonies numerous times, and each time the marmots have been nonchalantly scurrying around my tent within an hour or two.

One summer in the Siffleur Wilderness Area, I had a horrible time trying to photograph a big old male marmot. He kept wanting to come and visit me, licking my boots and tripod, and even chewing on my socks (while I had them on!).

Marmots are not always so curious, diving into their burrows en masse at the first hint of real danger, such as a golden eagle soaring high above. Other predators include grizzly bears, coyotes and wolverines. Marmots are the largest squirrels in the Canadian Rockies--males weigh 5-6 kg (11-13 lb) and are almost 1 m (3 ft) in length (females are significantly smaller).

Mike Potter

Hoary marmots are so-called because of the off-white fluffy mantle of fur they have on their upper backs and shoulders when they emerge from hibernation in late spring. The rest of the coat, including the short bushy tail, is grizzled greyish brown with a rusty hue. Marmots are often mistaken for ground squirrels, but can be readily distinguished by their black feet and their body size (an adult marmot is the size of a cocker spaniel).

Marmots are gregarious rodents, preferring the company of other marmots. They often form large colonies, dotting alpine meadows with tens or hundreds of burrow entrances. Hoary marmots are grazers, eating just about anything near their colonies that is green. They spend most of their time feeding, play-fighting or relaxing, even during storms.

Female marmots give birth to four or five young every second spring. Because marmots hibernate for two-thirds of the year, it takes two full summers before the young reach full size and can be on their own, so females do not give birth every other spring, choosing instead to look after their half-grown litter.

Red Squirrel

Tamiasciurus hudsonicus

Terry Berezan

Anyone who has walked or hiked in the Canadian Rockies has probably been accosted by the small yet vocal red squirrel. Bold and defiant, the red squirrel will give a heart-felt scolding to anyone or anything that passes through its domain, even its enemies.

Characterized by a rusty red-brown coat and an off-white yellowish underbelly, red squirrels have large bushy tails (often held along their backs) and white crescents around the eyes. Both sexes weigh about 200 g (7 oz) and measure 30 cm (12 in) in total length, including the 12 cm (5 in) long tail.

Red squirrels are primarily arboreal (tree-dwelling). They feed extensively on the seeds of pine and spruce trees, eating the seeds on the cones like we eat corn on a cob. They also eat mushrooms, nuts, seeds from other coniferous trees, berries, and the occasional mouse or stash of bird eggs.

Squirrels are favourite prey of martens, fishers, hawks and owls. Martens and fishers are actually faster and more agile than squirrels, on land or in the trees. Fortunately for squirrels, they are prolific breeders, with females giving birth each year to as many as eight blind and naked young in an underground nest.

Red squirrels have small home ranges and are solitary individuals except during the short breeding season. They frequent certain feeding areas year-round, resulting in large piles of discarded seed scales accumulating slowly over time. These old seed piles, called middens, are used by the squirrels as rest stops or as shelter (they burrow in) for short spells in winter when the temperature dips dramatically. Unlike most members of the squirrel family, red squirrels do not hibernate, venturing out in mild conditions. During cold snaps, they feed on cones harvested in the fall and stored on top of the midden but below the snow.

Red squirrels are abundant in coniferous forests throughout the Central Rockies. They are active during daylight hours and can be spotted in all seasons.

Golden-mantled Ground Squirrel

Spermophilus lateralis

Al Williams

Golden-mantled ground squirrels prefer rocky country, usually in the subalpine or alpine zones. On a hike one summer to Sentinel Pass, the highest trail-accessible pass in Banff National Park, I was surprised to see two of these little ground squirrels right in the barren rock-filled pass.

Golden-mantled ground squirrels are about half the weight of Columbian ground squirrels, and look like oversized chipmunks. The biggest difference between a golden-mantled ground squirrel and a least chipmunk is that the body stripes on a chipmunk extend from the base of its tail to its nose, while on the ground squirrel they stop at the shoulders and don't carry on to the neck and head.

These ground squirrels have rusty, light brown heads (the golden mantles of the common name) and white undersides. Like other squirrels in the Central Rockies, they have white crescents around their eyes, but their tail is shorter and less bushy. Adult golden-mantled ground squirrels tip the scales at just over 200 g (7 oz) and are 30 cm (12 in) in total length, including the 10 cm (4 in) long tail.

Golden-mantled ground squirrels do hibernate, but intermittently wake up to feed and urinate. They are preyed upon by a host of predators, including hawks, coyotes and weasels. However, like most rodents, they can sustain a high rate of predation because they produce large litters — in this case, 4-6 young each spring.

Flowers, seeds, leaves, grasses and fungi make up the majority of a golden-mantled ground squirrel's diet. Gregarious and social, they often live in groups and will even live together with marmots or chipmunks. They are abundant in the Central Rockies, but are seen less often than the other members of the squirrel family because of the high altitude terrain they live in.

Columbian Ground Squirrel

Spermophilus columbianus

John Marriott

Columbian ground squirrels are the most common grassland squirrel in the Central Rockies, frequenting open meadows at most elevations from the montane to the alpine.

In contrast to their prairie cousins, the Richardson's ground squirrel or "gopher," Columbian ground squirrels are richly coloured, with black and white speckled backs tinged honey brown, and cinnamon-coloured noses, chests and undersides. They have pale white crescents around each eye, and a bushy black and white speckled tail that is also tinged brown.

Highly sociable rodents, Columbian ground squirrels live in colonies that are divided into separate territories ruled by dominant males. Each colony features a network of burrows and dens, all cleverly designed with multiple entrances and escape hatches and built so that the dens stay high and dry through even the worst of storms. Sentries are constantly on the watch for danger and when a threat is spotted will signal a warning to other members of the colony by giving a shrill, piercing whistle.

An adult Columbian ground squirrel is about 30 cm (1 ft) long and weighs at most 500 g (1.1 lb). Although fairly small, Columbian ground squirrels represent tasty treats for a number of predators, in-

Mike Potter

cluding coyotes, weasels, badgers, foxes, hawks and grizzly bears (which dig the ground squirrels out of their hibernation dens).

Mating season occurs in June, with males battling it out to win the right to breed. Female Columbian ground squirrels give birth to litters of up to 8 young (average 4) a month later. Once the young are weaned (another month later), they begin grazing like their parents, enjoying a natural diet of roots, stems and leaves from a variety of flowers and plants.

Columbian ground squirrels are easily approached by humans, particularly in areas where they have been fed by hand. However, they are not the cute little creatures they may seem when they become used to our presence, occasionally biting and injuring people. In any case, their digestive systems do not handle human food very well - it may cause them illness or even death.

Columbian ground squirrels are readily spotted scurrying about in most picnic areas and meadows in the Central Rockies during the summer months. They hibernate from September to April at low elevations and from October to early June in the high country.

Least Chipmunk

Eutamias minimus

Mike Potter

The least chipmunk is the smallest member of the squirrel family in the Central Rockies. Least chipmunks are smaller than the similar-looking golden-mantled ground squirrels, and their distinctive back stripes run from the rump right to the tip of the nose. They have five black stripes, with four off-white stripes in between. The entire coat of the least chipmunk is tinged rusty brown, with the colour especially pronounced on the sides.

Female least chipmunks are slightly larger than their male counterparts, weighing approximately 50 g (2 oz) and measuring 22 cm (9 in) long, including a 10 cm long tail. Least chipmunks begin hibernating in October and emerge in April, awakening occasionally during the winter to feed on stored supplies. Mating occurs soon after they appear in the spring, resulting in litters of 4-5 young each summer. Weasels, hawks, owls and coyotes prey on both young and adults.

With little paws that act like human hands, least chipmunks will often stuff their cheeks full of their favourite seeds and retreat to shelter before commencing eating. They seem to have a nervous energy abounding, for they are usually spotted scurrying about with their tail up. Least chipmunks, with their "racing stripes," are common throughout the Central Rockies and can often be seen near creeks and streams, in open forests and on rocky ledges.

Muskrat

Ondatra zibethicus

Al Williams

The muskrat is one of the more common and visible mammals in the waters of the Central Rockies. Named for two anal glands that emit a musk used to mark territories and entice mates, these medium-sized rodents are more closely related to voles than they are to rats.

Muskrats are often mistaken for beavers, since both are semi-aquatic and have dark brown coats. However, muskrats are considerably smaller than beavers and have long round tails compared to the broad flat tails of beaver. At a distance in the water, a muskrat can be distinguished from a beaver by the way its tail waves back and forth leaving ripples in its wake (a beaver's tail doesn't wave back and forth and leaves behind a smooth wake).

An adult muskrat is chunky in appearance, but despite being 50 cm (20 in) long, weighs only about 1 kg (2.2 lb). Other than its pinkish webbed feet and dark tail, the body is covered in buoyant fur that is one of the muskrat's many special adaptations for aquatic life. Like beavers, muskrats have large front incisors with the lips in behind so that they can chop plant stalks and tubers without getting water in their mouths. They are also equipped with special muscles that close their eyes and nostrils automatically when they dive. An average dive for food or to escape an enemy may last from 3-5 minutes, although muskrats have enormous lung capacities that enable dives of up to 15 minutes.

Muskrats eat primarily aquatic plants, but are omnivorous and will also eat a variety of frogs and fish. After catching or snipping their meal, they move on to a feeding platform (constructed from plant leaves and mud) to eat and groom themselves. Each spring, I have seen muskrats also using the receding ice on ponds and lakes as temporary feeding platforms.

Muskrat dens are usually burrows carved out of pondside banks or small lodges constructed in the lake or pond from vegetation. Muskrats do not hibernate and do not store up food for the long Rockies winters, so they also construct several small lodges or "push-ups" of vegetation in their territory. There they can take warm breathing stops while foraging under the ice.

Muskrats are territorial, fighting whenever challenged. They are particularly ferocious during the spring breeding season, with battles between males often ending in injury or even death. The winners pair off for a year with a female, which is impregnated in April or May and gives birth a month later to a litter of up to eight young. There is usually another litter of the same size later in the summer.

This prolific breeding is kept in check due to regular hunts by mink, hawks, owls, coyotes and other predators.

Muskrats, though mainly nocturnal, are active around the clock. They can be spotted in ponds, near the edges of small lakes such as Johnson Lake in Banff National Park, and in slow-moving waterways.

Bushy-tailed Woodrat

Neotoma cinerea

Paul Smith

Bushy-tailed woodrats are generally a light greyish or tawny brown, with black streaks or markings on top and pale undersides. They have big ears and eyes like mice, and a squirrel-like tail that is often as long as the body (overall length is about 40 cm (16 in), weight up to 400 g (14 oz)).

Bushy-tailed woodrats are excellent climbers, frequenting coniferous forests, where they nest under boulders by preference, while also using caves, rocky outcroppings, cliffs and old buildings. Although they are common and active year-round, their droppings are seen far more often than they are because of their nocturnal nature.

Bushy-tailed woodrats' insatiable habit of collecting and hoarding things, including smooth stones, feathers and shiny objects, has earned them the nickname of "packrats." More than once they have been known to steal a human belonging or two left unattended in the backcountry. They will even trade an object they are carrying for a more attractive one they come upon.

The woodrat's primary diet is the foliage of deciduous trees; they also eat fruits, seeds and coniferous foliage. Predators include hawks, owls, weasels and coyotes.

Bushy-tailed woodrats do not hibernate, but do stockpile food in their den for the winter months. Breeding occurs in mid-winter, with raucous fighting among the males. A month later, 3-4 young are born. A second litter often occurs later that spring.

Deer Mouse

Peromyscus maniculatus

Paul Smith

On a short jaunt late at night one summer while camping in Kootenay National Park, my flashlight picked up a little mouse sprinting along in front of me, stopping every few feet to glance back at me. It was no surprise to see a deer mouse out under the moonlight, as they are strictly nocturnal. After a while of sitting on a log in silence, I was able to hear and spot a couple of them moving about collecting seeds in the underbrush.

Deer mice range in colour from light to dark brown, with pale bellies, white feet, large eyes and ears for their size, and a long thin tail coloured brown on top, white underneath. Adults are usually 15-20 cm (6-8 in) long, with the tail representing almost half that length. Average weight of a deer mouse is 30 g (1 oz).

Like most other mice, deer mice are active breeders, becoming sexually mature five or six weeks after birth and having 2-3 litters of 1-8 (average 4) young in a summer. Their population is regulated by a variety of terrestrial and aerial night hunters including owls, coyotes and weasels.

Active year-round, deer mice are very common throughout the Central Rockies at just about any elevation. They feed mostly on seeds, storing them in safe caches for colder weather, but will also eat other foods like berries and insects.

Snowshoe Hare
Lepus americanus

Al Williams

Snowshoe hares, also called varying hares, may be the best camouflage artists in the Canadian wilds. On an extended backpacking trip one summer, I had set up my tent and was preparing dinner when I spotted a movement off to the right. I stared at the spot but couldn't see anything. Finally, I walked over, and nearly died of shock when a snowshoe hare leapt out from under me and bounded off to safety.

The grey-brown coat that helps snowshoe hares disappear into their surroundings is a major part of their protective gear. In winter, when their coat turns to snow white except for faint black ear tips, they again become a part of the landscape, blending perfectly against the backdrop. Snowshoe hares are also equipped with blazing speed and agility, and long snowshoe-like hind feet that enable them to fly across deep snow away from owls, coyotes, wolves, fishers, and their main predator, the lynx. In winter, it is easy to spot the runs of snowshoe hares because of their distinctive tracks - the large hind feet always appear ahead of the small front feet.

Snowshoe hare and lynx populations are intricately linked in a periodic cycle, with lynx numbers fluctuating with the number of snowshoe hares. The hare population rises gradually over an eight to ten year period until it reaches a peak where there are too many

Paul Smith

hares for the habitat to support, causing a population crash and a similar later decline in the lynx population.

In the summer, snowshoe hares are primarily grazers, eating grasses and forbs, while in winter they become browsers, relying on bark and shrubbery to fill their dietary needs. Adult female hares, called does, weigh 1.5 kg (3.3 lb) on average and are about 46 cm (18 in) long, including the small puffy tail. The males, or bucks, are slightly smaller.

Breeding begins in late winter and continues through the spring. In May, a litter of up to 7 leverets (average 3-4) is born, and in the summer each doe usually gives birth to one more litter (occasionally two more). The baby hares are born furry and ready for action, and are weaned within a week. However, only a small percentage make it to their second year due to predation, disease and starvation.

Snowshoe hares are primarily nocturnal, although they are sometimes active at dusk or dawn. During daylight hours, they normally rest in a "form" - a beaten-down, well-hidden spot beneath a tree branch or in a cluster of shrubs and tall grasses. Snowshoe hares are forest dwellers and frequent areas that support dense young growth.

Pika

Ochotona princeps

Mike Potter

High in the alpine, amidst boulder-strewn fields and slopes, lives one of the Rockies' most intriguing mammals, the pika (pronounced "pee-ka"). This little rock rabbit looks more like a miniature guinea pig with big ears than a relative of the common rabbit.

Whenever a predator or danger is spotted, pikas shriek a high-pitched "eeep" as a warning call. They use a slightly different pitch to declare their territory to other pikas.

Pikas are light orangish-brown, with grizzled touches of black, grey and dark brown on their backs to help them blend into their rocky homes. They do not have a visible tail, but do have large round ears fringed in white. Adult pikas are small yet stocky, measuring 18 cm (7 in) long and weighing 190 g (6.5 oz).

Pikas eat grasses, greens and lichens, and spend a good part of each summer day stockpiling vegetation in clumps all over the rocks to let it dry out. Once the "hay" is dry, the pika will take it and store it in the burrow beneath the rocks in preparation for the long winter ahead. This food cache is a valuable resource and is essential to pikas' survival in winter, since they do not hibernate. Though they live in loosely associated colonies, pikas are territorial and will resolutely defend both their home range and their food cache.

Like other members of the rabbit order, pikas will eat special partially digested pellets. They do not reingest normal feces.

In winter, pikas live in the network of paths they establish beneath the rocks and snow. The crevices are handy in summer to escape predators like the golden eagle, but are not as helpful against their main predators, the smaller weasels, which are small and agile enough to be able to go anywhere the pika goes.

Pikas produce litters of 3-5 young each spring. Second litters later in the summer are common, but few of the second batch survive because of the short interval they have to gather a food cache for the upcoming winter. Like baby snowshoe hares, pika young are precocious, born furry and mobile.

Although their calls are ventriloquistic, pikas are readily spotted with a little patience when in talus in the alpine zone.

Paul Smith

Red Fox *Vulpes vulpes*

In places where coyotes and wolves coexist, red foxes are rare, and such is the case in the Central Rockies. This small wild dog can be recognized by its long bushy tail and rusty orange coat. Red foxes are occasionally spotted in Kananaskis Country, but sightings in other parts of the Central Rockies are extremely rare.

Bobcat *Felis rufus*

Smaller than a lynx, with the appearance of an oversized house-cat with ear tufts, the bobcat is a shy, elusive feline that mainly hunts rodents, hares and birds, as well as deer once in a while.

Bobcats are seldom seen even in places where they are common, and their distribution does not include the Central Rockies Ecosystem except for the possibility of a few individuals along the western slopes and in the foothills.

Fisher *Martes pennanti*

Members of the weasel family, fishers are larger than their close cousin, the marten. A rich brown fur covers the fisher, which is nocturnal and is one of the few carnivores that has devised a successful gameplan for killing porcupines, a staple of its diet. The fisher attacks the unprotected nose of the porky repeatedly, then, when it gets the chance, flips the porcupine over onto its back to expose the quill-less underbelly.

Fishers are solitary, reclusive animals that occur at low densities in dense subalpine forests. They are present throughout the Central Rockies Ecosystem, but are rarely sighted because of their secretive nature and inhospitable home terrain.

River Otter *Lutra canadensis*

Long, slim and tapered like many other members of the weasel family, river otters are more at home in water than on land. These dark brown carnivores love to play, and will toboggan down mud or snow slopes on their bellies. Excellent swimmers and divers, they will eat rodents (including beavers), fish, frogs, birds and eggs.

River otters are found in low densities in Kananaskis Country, but are rare elsewhere in the Central Rockies.

Badger *Taxidea taxus*

Badgers are digging machines, using their flattened bodies, powerful legs and long sharp claws to dig for their food. They eat ground squirrels and just anything else they can find above or below ground.

This large member of the weasel family is uncommon in the Central Rockies, save for the eastern foothills in Kananaskis Country and dry low-lying valleys on the east side of Banff National Park.

Richardson's Ground Squirrel *Spermophilus richardsonii*

The Richardson's ground squirrel closely resembles the Columbian ground squirrel, but is paler in colour without the orange tinge, does not have as bushy a tail, and does not have the white eye crescents. Richardson's ground squirrels are primarily a prairie species, only found in low numbers in the foothills in the southeastern part of the Central Rockies.

Northern Flying Squirrel *Glaucomys sabrinus*

While this "flying" rodent is actually common in the Central Rockies, it is so rarely seen as to merit only an inclusion in this appendix. Strictly nocturnal, these squirrels (slightly larger than red squirrels) have patagia, loose flaps of skin on their sides connecting their front and back legs, that allow them to glide from tree to tree.

Northern flying squirrels are active at night and can be seen by moonlight in openly spaced coniferous forests. They are omnivorous, and will eat meat as well as seeds and berries.

Other Mammals

The Central Rockies Ecosystem is home to a variety of small mammals other than those described in this book. These include several species of voles, such as meadow vole and red-backed vole, that are important food sources for many of the smaller carnivores; several highly insectivorous shrew species; and a number of bat species, including the fairly common little brown bat.

Alpine zone The land above the trees, including rocks, meadows and glaciers.

Carnivore Meat-eating mammal; member of the order Carnivora.
Central Rockies Ecosystem The area encompassing the foothills, front ranges, and main ranges of the Canadian Rocky Mountains between the Crown of the Continent Ecosystem to the south and the Greater Jasper Ecosystem to the north. Includes protected areas from Kananaskis Country in Alberta and the Height-of-the-Rockies Wilderness Area in British Columbia north to Banff National Park in Alberta and Yoho National Park in B.C. (see map p. 6).
Coniferous Trees that reproduce by means of cones; conifers have needles (modified leaves) that are usually evergreen.

Deciduous Trees or plants with leaves that fall off each year before winter.

Family A group of related organisms, containing one or more genera, in the system of biological classification of all living things.

Genus (plural genera) A group within a family, containing one or more closely related species.
Gregarious Living together in groups.

Habitat The particular area or type of area required for an organism, influenced for mammals by factors such as temperature, elevation, ground cover, plant growth and the presence of other species of mammals.
Harem A herd of females gathered for mating purposes by one male.
Herbivore Plant-eating mammal; member of the order Herbivora.
Hibernate To spend periods of time (usually the winter) in a torpid state, with bodily functions slowing down and body temperature dropping below normal.
Home range The area that an individual animal uses.

Mammal A member of Mammalia, the highest class of vertebrates. Distinguishing characteristics shared by mammals include having highly developed brains and senses, being warm-blooded, having hair and having nipples capable of producing milk.
Montane zone The low-elevation vegetation zone below the sub-alpine zone, above which trembling aspen does not grow.

Nocturnal Active at night.

Omnivore A creature that eats both plant and animal matter.

Pelage The fur or coat of an animal.
Precocious Young that are born highly developed, requiring little or no care before being fully functional.

Rodent Small gnawing mammals with two pairs of large front incisor teeth; member of the order Rodentia.
Rut The period of sexual excitement in which males seek out females to mate with them.

Species An individual member of a genus.
Subalpine zone The vegetation zone between the montane and alpine zones, characterized by coniferous trees.

Ungulate A hoofed mammal.

Wildlife corridor A travel route utilized by wildlife to move from one area to another.

Alberta Wildlife Viewing Guide, 1990, Lone Pine Publishing, Edmonton. Excellent tips on how and where to view wildlife in Alberta.

Banfield, A.W.F., *The Mammals of Canada,* 1974, University of Toronto Press, Toronto. Very technical descriptions, in-depth ecological information, Canada-wide.

Forsyth, Adrian, *Mammals of the Canadian Wilds,* 1985, Camden House Publishing Ltd., Camden East, Ontario. Photographs, descriptions, Canada-wide.

Gadd, Ben, *Handbook of the Canadian Rockies, Second edition,* 1995, Corax Press, Jasper. Excellent descriptions and anecdotes, illustrations.

Hummel, Monte and Sherry Pettigrew, *Wild Hunters - Predators in Peril,* 1991, Key Porter Books Ltd., Toronto. Good coverage of large carnivores, Canada-wide focus.

Lynch, Wayne, *Wildlife of the Canadian Rockies,* 1995, Alpine Book Peddlers, Canmore. Coffee table photographic book, portraits.

Scotter, George W. and Tom J. Ulrich, *Mammals of the Canadian Rockies,* 1995, Fifth House Ltd., Saskatoon. Excellent photographs, comprehensive descriptions, coverage of all species.

Van Tighem, Kevin, *Wild Animals of Western Canada,* 1992, Altitude Publishing Ltd., Canmore. Photographs, descriptions, personal anecdotes.

The author photographing in alpine meadows near Dolomite Pass in Banff National Park. Photo: Jeff Waugh

John Marriott began working as a park naturalist in Banff National Park in 1992. His interest in wildlife photography and writing blossomed into a part-time profession in 1994 after he graduated from the University of British Columbia with a Bachelor of Science degree in Forestry specializing in Parks Management.

John's photographs have been published in *Canadian Geographic* magazine, on the *Canadian Rockies & Wildlife* CD-ROM, and on the *Canadian Rockies Net* website.

JEM Photography, John's business, offers guided photographic tours, photography workshops and photography trip planning/consulting. His home page on the Internet is at:

www.canadianrockies.net/Jem_Photography/

John and his wife, Christine, live in Canmore, Alberta, just outside of Banff National Park.